An Introduction to the Smarandache Function

Charles Ashbacher

Decisionmark
200 2nd Ave. SE
Cedar Rapids, IA 52401 USA

Erhus University Press
Vail
1995

The graphs on the covers belong to Dr. I. Balacenoiu, "The Monotony of Smarandache Functions of First Kind" (to appear in 1996).

"An Introduction to the Smarandache Function",
    by Charles Ashbacher

*Copyright 1995 by Charles Ashbacher
    and Erhus University Press*

ISBN 1-879585-49-9

Standard Address Number 297-5092

Printed in the United States of America

Preface

The creation of a book is an act that requires several preconditions.

1) An interesting and worthwhile subject.

2) A fair, yet demanding editor.

3) Someone willing to put the words on paper.

Given the existence of these three items, the leap to making the book a good one becomes the responsibility of the author. The Smarandache function is simultaneously a logical extension to earlier functions in number theory as well as a key to many future paths of exploration. As such, it is hoped that you get as much out of this book as the author did in writing it. While all books are a collective work, the responsibility for any errors ultimately falls to the author, and this is no exception.

Several conjectures are made in this book and while the author believes that they are true, there will be no offence taken if any are proven wrong. Progress in mathematics is often made by seeing such opinions and proving them wrong. In fact, it is hoped that if any reader makes any progress in resolving any conjecture, that they will have the good grace to contact the author at the address below.

I would like to take this opportunity to thank Dr. R. Muller for his encouragement in getting this project started. The staff at Erhus University Press are also to be commended for putting everything in the proper place. Thanks also must go to my supervisors Toufic Moubarak and Brian Dalziel for their understanding during the creation of the book.

Finally, I would like to dedicate this book to my lovely daughter Katrina. A model of chaos theory, she is capable of creating messes that make a tornado look like a sneeze. And yet she somehow manages to stay cute while doing so.

Charles Ashbacher
Decisionmark
200 2nd Ave. SE
Cedar Rapids, IA 52401 USA
October, 1995

3

Chapter 1

As one of the oldest of mathematical disciplines, the roots of number theory extend back into antiquity. Problems are often easy to state, but extremely difficult to solve. Which is the origin of much of their charm. All mathematicians, amateurs and professionals alike, have a soft spot in their hearts for the "purity" of the integers. When "Fermat's Last Theorem" was finally proven after centuries of effort, the result was discussed on many major news shows in the US. Brief comments also appeared in the major weekly news magazines.

Divisibility is the backbone of number theory. Notions such as prime numbers and the standard number theoretic functions

The Euler phi function, $\Phi(n)$ is the number of integers m, where $1 \leq m \leq n$ and m and n are relatively prime

Sum of divisors function, $\sigma(n)$ is the sum of all the positive divisors of n

Number of divisors function, $\tau(n)$ is the number of positive divisors of n.

are all based on which integers are evenly divisible by others.

Divisibility and prime factorizations are intimately related, therefore the values of number theoretic functions can often be computed by formulas based on the prime factorization.

For example, if $m = p_1^{\alpha 1} \ p_2^{\alpha 2} \ p_3^{\alpha 3} \ \cdots \ p_n^{\alpha n}$, then

$$\Phi(m) = m\left(1 - 1/p_1\right)\left(1 - 1/p_2\right)\left(1 - 1/p_3\right) \cdots \left(1 - 1/p_n\right)$$

In the 1970's a Rumanian mathematician by the name of Florentin Smarandache created a new function in number theory. Called the Smarandache function in his honor but not published until 1980[1], it also has a simple definition

If $m > 0$, then $S(m) = n$, where n is the smallest number ≥ 0 such that m evenly divides n!.

It is possible to extend the domain of S to include the negative integers as well. However, since the basic notions of divisibility also includes the negative integers, i.e.

If $m \mid n$, then $-m \mid n$

4

nothing really new occurs if this is done. Do note however, that if the domain is expanded in this way,

$$S(m) = S(-m)$$

so S is an even function. Of course, zero is always excluded. For all of our work here, we will consider the domain to be m > 0.

Some example values are

S(1) = 0 since   0! = 1
S(2) = 2 since   0! = 1, 1! = 1 and 2! = 2
S(3) = 3 since 3 does not divide 0! = 1, 1! = 1 or 2! = 2, but 3 divides 3! = 6
S(6) = 3 since 6 does not divide 0! = 1, 1! = 1 or 2! = 2, but 6 divides 3! = 6
S(16) = 8 since 16 divides 8! and no other factorials less than that

Extending out to an infinite horizon, the consequences of this simple definition encompass many areas of mathematics. Sometimes behaving like the standard functions of number theory and other times totally different, this function occupies a unique niche all its' own. Our purpose here is to examine some of those consequences, hopefully giving the reader an acquired taste for this unexplored territory. In all of our explorations, the reader should assume that all numbers are non-negative integers unless otherwise noted. Also, the word divides means evenly divisible and for notational consistency p and q will always be used to denote primes. Finally, one will not be considered a prime number.

We start our journey in a simple manner, with a table of the values of the function for the first 30 numbers.

Table 1

n	S(n)	n	S(n)	n	S(n)
1	0	11	11	21	7
2	2	12	4	22	11
3	3	13	13	23	23
4	4	14	7	24	4
5	5	15	5	25	10
6	3	16	6	26	13
7	7	17	17	27	9
8	4	18	6	28	7
9	6	19	19	29	29
10	5	20	5	30	5

Note that $S(3) = S(6) = 3$, so $S(n)$ is not 1-1. This behavior is typical of number theoretic functions.

Looking at table 1, it can be seen that $S(n) \leq n$ for all 30 entries. This is a general result and is the topic of our first theorem.

Theorem 1: $S(n) \leq n$ for all $n \geq 1$.

Proof: Choose an arbitrary $n \geq 1$ and construct the list

$$0, 1, 2, 3, \ldots n$$

with corresponding factorials

$$0!, 1!, 2!, 3!, \ldots, n!$$

since we are looking for the smallest number m in this list such that n divides $m!$, we start at the left and move right until we encounter such a number. Since n divides $n!$ for all $n \geq 1$, in all cases we need go no further than n.

Contrasting this with the standard number theoretic functions

$$\Phi(n) \leq n \qquad \sigma(n) > n \qquad \tau(n) < n$$

we can see that this is a case where the behavior of S is similar.

Clearly,

$$\sum_{k=1}^{\infty} 1/S(n) \quad \text{diverges}$$

since

$$\frac{1}{S(n)} \geq \frac{1}{n}$$

for all n and the harmonic series

$$\sum_{n=1}^{\infty} 1/n$$

is known to be divergent.

Our second theorem gives a lower bound to the values.

Theorem 2: $S(n) > 1$ for $n \geq 2$

Proof: Given that $0!$ and $1!$ both equal 1 and any number $n \geq 2$ has at least one prime factor greater than or equal to 2, it follows that the smallest number m such that $n \mid m!$ must be greater than 1.

Corollary 1: $0 \leq S(n)/n \leq 1$ for all $n \geq 1$.

Proof: Since the domain of the function is all positive numbers, it is a direct consequence of theorem 1 that

$$S(n)/n \leq 1$$

Combining the results of theorems 1 and 2, we have that $S(n)/n > 0$ for $n > 1$. Including the special case $S(1) = 0$ yields $S(n)/n \geq 0$

There are 10 prime numbers less than 30, and in looking at table 1, we see that $S(n) = n$ for all 10 prime numbers. This is a general result and easy to prove.

Theorem 3: $S(p) = p$ for p a prime.

Proof: Choose an arbitrary prime number p and construct the list of factorials

$$0!, 1!, 2!, \ldots, p!$$

To determine the value of $S(p)$ we start at the left and move right until we encounter the first number m where p divides $m!$. By the definition of the primes, there is no number less than p that contains p as a factor. Therefore, p is the smallest number where p divides $p!$.

This is similar to the behavior of two other number theoretic functions in that

$$\Phi(p) = p - 1 \text{ for p a prime}$$

$$\sigma(p) = p + 1 \text{ for p a prime}$$

The number of divisors function is different, in that

$$\tau(p) = 2 \text{ for p a prime}$$

With this result, we also know that the series

$$\sum_{k=1}^{\infty} 1 / S(p_k)$$

7

where p_k is the k-th prime diverges since the sum of the reciprocals of the primes is known to diverge.

Corollary 2: The equation $S(n)/n = 1$ has an infinite number of solutions.

Proof: Direct consequence of theorem 3 and the well-known fact that there is an infinite number of prime numbers.

Going back to table 1 once again, we see that

$S(6) = S(2*3) = 3 = S(3)$

$S(15) = S(3*5) = 5 = S(5)$

$S(26) = S(2*13) = 13 = S(13)$

In all three cases the input number is the product of two distinct primes and the value of the function is the largest of the primes. This is again no coincidence and is the subject of the next theorem.

To complete the proof we rely on the following property of factorials.

Fact: If m | k!, then m | n! for all n > k.

Theorem 4: If p and q are distinct primes, then S(pq) = largest of p and q.

Proof: Without loss of generality, assume that p > q. Form the list of factorials

$0!, 1!, 2!, \ldots, q!, \ldots, p!$

Since p is prime, as we move from left to right in this list, the first factorial we encounter that is divisible by p is p!. Since q also divides p!, it follows that pq divides p!. Since this is the smallest such number, p satisfies the definition of the Smarandache function.

Definition 1: A function f is said to be multiplicative if

$f(m*n) = f(m) * f(n)$ for all m,n in the domain of f.

Corollary 3: S(m) is not multiplicative.

Proof: Direct consequence of theorem 4.

Since σ, τ and Φ are all multiplicative, this is a fundamental point of difference between S

8

and other number theoretic functions..

However, given that $S(2*2) = 4 = S(2)*S(2)$, it is possible to find numbers m and n such that $S(mn) = S(m)*S(n)$.

As is proven in the following theorem, the behavior described in theorem 4 can be generalized.

Theorem 5: Let $m = p_1 p_2 \ldots p_k$, where all p_i are distinct primes. Then,

$$S(m) = \text{largest of the set } \{ p_1, p_2, \ldots, p_k \}.$$

Proof: First reindex the set so that the order of the indices matches the order of the primes. Form the ordered list of factorials

$$0!, 1!, \ldots, p_1!, \ldots, p_2!, \ldots, p_k!$$

and apply the same reasoning used in theorem 4.

Definition 2: We will use the notation $NS(m) = n$ to denote that n is the number of different integers k such that $S(k) = n$.

Example:
$NS(0) = 1$ since $S(1) = 0$ and there are no other numbers n such that $S(n) = 0$

The order of the growth of $NS(m)$ will at times be of importance, so we start here with a straightforward consequence of theorem 5.

Theorem 6: If we restrict the domain of $NS(p)$ to be the odd primes only, then the growth in the value of $NS(p)$ is at least

$$\sum_{n=0}^{\pi(p)-1} \binom{\pi(p)-1}{n}$$

where the function $\pi(x)$ is the number of prime numbers not exceeding x and

$$\binom{m}{n} = \frac{m!}{n!(m-n)!}$$

Proof: Let p be an arbitrary prime. By theorem 5, any number of the form $n = kp$, where k is a product of distinct primes each less than p will satisfy the equation

$$S(kp) = S(p) = p$$

There are $\pi(p)$ -1 primes less than p. The number of ways we can create a product by choosing k items from this list is

$$\binom{\pi(p)-1}{k} \qquad 0 \leq k \leq \pi(p)-1$$

Since each product is distinct, the total number of ways we can create the products is the sum of these values for all possible choices for k.

Example: If we choose $p = 31$, then $\pi(31) - 1 = 10$. By the theorem, NS(31) is at least

$$\binom{10}{0} + \binom{10}{1} + \binom{10}{2} + \binom{10}{3} + \binom{10}{4} + \binom{10}{5} + \binom{10}{6} + \binom{10}{7} + \binom{10}{8} +$$

$$\binom{10}{9} + \binom{10}{10} =$$

$1 + 10 + 45 + 120 + 210 + 252 + 210 + 120 + 45 + 10 + 1 = 1024$

Moving up to the next prime $p = 37$, $\pi(37) - 1 = 11$, so the corresponding number is

$1 + 11 + 55 + 165 + 330 + 462 + 462 + 330 + 165 + 55 + 11 + 1 = 2048$

Which is an illustration of the following well-known theorem concerning such sums.

Counting Principle 1: For $n \geq 0$

$$\sum_{k=0}^{n} \binom{n}{k} = 2^n$$

Since this result is well-known the proof is not given here. Interested readers may consult any book on combinatorics for the details.

Corollary 4: For p any prime, the value of NS(p) is at least

$$2^{\pi(p)-1}$$

Proof: Direct consequence of theorem 6 and counting principle 1.

Example:
There are 88 primes less than 1000, so for $p = 1009$ we know that there are at least

$$2^{88}$$

numbers n where $S(n) = 1009$.

10

And, as we will see later, this is only part of the growth of NS(p) for p "sufficiently large."

The range of the ratio S(n)/n has already been determined. However, it can now be proven that the ratio can be made arbitrarily close to zero.

Theorem 7: For any arbitrary real number $\epsilon > 0$, it is possible to find a number n such that S(n)/n < ϵ.

Proof: Choose any real $\epsilon > 0$. Form a product of distinct primes q = $p_1 p_2$ . .p_k such that 1/q < ϵ. Now, take another prime s that is larger than all of the prime factors of q and form the product sq. By theorem 5, S(sq) = s, so the ratio S(sq)/sq = 1/q.

Since S(p) = p, for p a prime, it is clear that

$$\sum_{k=1}^{\infty} S(k)/k \quad \text{diverges}$$

However, many sums can be made to converge by selectively eliminating the part that diverges. So, suppose we eliminate the primes and ask the related question:

What is the behavior of the sum

$$\sum_{k=4}^{\infty} S(k)/k \quad \text{where k ranges over all composite numbers}$$

To see that this sum also diverges, consider all numbers m, where m = 2p, p an odd prime. It follows from theorem 4 that

$$S(m)/m = p/2p = 1/2$$

and since there are an infinite number of odd primes, the sum must diverge.

It has already been determined that

$$\sum_{n=1}^{\infty} 1/S(n) \quad \text{and} \quad \sum_{k=1}^{\infty} 1/S(p_k) \quad \text{where } p_k \text{ is the kth prime}$$

diverge. At this time, it will be proven that the sum

$$\sum_{n=4}^{\infty} 1/S(n) \quad \text{where n is composite}$$

11

also diverges.

Consider all composite numbers of the form n = 2p, where p is an odd prime. By theorem 4,

$$S(2p) = p$$

And so the sum

$$\sum_{k=1}^{\infty} 1/S(2p_k) \quad \text{where } p_k \text{ is the kth odd prime}$$

is a divergent subseries of

$$\sum_{k=4}^{\infty} 1/S(n) \quad \text{where n is composite}$$

forcing it to also be divergent.

The next step in our journey is to consider the behavior of S(m) for increasing powers of a prime. We will start with the simplest case, using the proof as a building block to solve several generalizations.

Definition 3: If p is a prime, we define the function $N_p(m)$ to be the number of instances that p is a factor of m!.

Lemma 1: The behavior of the function $N_p(m)$ is given by the formula

$N_p(m) = N_p(kp)$ for kp ⁂ m < (k+1)p

$N_p(m) = N_p(kp) + j$ for m = (k+1)p

where j is an integer   1.

Proof: Form the series of numbers

kp, kp+1, kp+2, . . . , (k+1)p - 1, (k+1)p

and associated factorials

kp!, (kp+1)!, . . . , ((k+1)p - 1)!, ((k+1)p)!

It is a direct consequence of the definition of prime numbers that p does not divide any of

12

the numbers

$$kp+1, kp+2, \ldots, (k+1)p - 1$$

Therefore, if we split the factorial $((k+1)p - 1)!$ up into two components

$$(1*2*3* \ldots *kp) * ((kp+1)*(kp+2)* \ldots *((k+1)p-1))$$

p does not divide the second component. Therefore, the number of instances of p as a factor of any number

$$(kp+1)! \quad \text{up to} \quad ((k+1)p-1)!$$

must be the same as that of kp!

Now taking $((k+1)p)!$ and splitting it into two components

$$((k+1)p-1)! * (k+1)p$$

it is clear that $(k+1)p$ will add at least one more instance of the prime p to the list of factors.

NOTE 1: If $p \mid (k+1)$, the factor $(k+1)p$ will add more than one instance of the prime p.

NOTE 2: $\lim_{n \to \infty} N_p(n) = \infty$

Theorem 8: If $m = p^2$, where p is a prime, then $S(m) = 2p$.

Proof: Again form the sequence of factorials

$$0!, 1!, 2!, \ldots, p!, (p+1)!, \ldots, (2p - 1)!, 2p!$$

and realize that for m to divide a number k, k must contain at least 2 instances of the prime p as factors. Starting at the left and moving right, the first number that contains one instance of p as a factor is p. So p! is the first factorial in the list evenly divisible by p. By lemma 1, all of the numbers $(p+1)!$ to $(2p-1)!$ also have only one instance of p as a factor. The additional required instance of p is not added until 2p is encountered. Since this is the smallest such number, we have the desired result.

Extending the arguments of the proof even further, we get the following general result.

Theorem 9: If $m = p^k$ where p is prime, then $S(m) = np$, where $n \leq k$.

Proof: The number we seek must contain at least k instances of p as a factor. By lemma 1, the number of instances of the prime in the factorials forms a stepwise function with increments only at the multiples of the prime p. Following the definition of the Smarandache function, since the smallest number containing at least k instances of the prime p is what we seek, that number must be an integral multiple of p. Since each multiple of p adds at least one instance of p, kp is the highest we ever need go.

The next step is to extend the ideas of the first note following the previous lemma. As an initial step we will form the factorials of the first few integral multiples of 3 and determine the number of times 3 appears as a factor.

Product	Number of 3's
1*2*3	1
1*2*3*4*5*(2*3)	2
1*2*3*4*5*(2*3)*7*8*(3*3)	4
9!*10*11*(3*4)	5
9!*10*11*(3*4)*13*14*(3*5)	6
15!*16*17*(2*3*3)	8

Where the corresponding values of the Smarandache function are

$S(3) = 3$ $S(3*3) = 6$ $S(3*3*3) = 9$ $S(3*3*3*3) = 9$ $S(3*3*3*3*3) = 12$
$S(3*3*3*3*3*3) = 15$ $S(3*3*3*3*3*3*3) = 18$ $S(3*3*3*3*3*3*3*3) = 18$

Pay special attention to the instances where two successive powers of three have the same value of the Smarandache function.

From this, it is easy to see that at least for the first few numbers m = 1, 2, 3, 4, 5, and 6, it is possible to find a power of 3 such that $S(3^k) = 3m$. The obvious question is then whether this holds in general. It does and is the topic of the next theorem.

Theorem 10: Let p be an arbitrary prime and $n \geq 1$. Then, it is possible to find a number k, such that

$$S(p^k) = np$$

Proof: We already know that $S(p) = p$ and $S(p^2) = 2p$. So assume that there is at least one number n > 2 such that there is no number k satisfying the equation

$$S(p^k) = np$$

Using the ordering properties of the natural numbers, there must be a smallest such n. For

14

notational purposes, call that number j. Therefore, there must be some number k such that

$$S(p^k) = (j-1)p$$

By the definition of S(m), the product

$$1*2*3*4* \ldots *(j-1)p$$

has k instances of the prime p and is the smallest such number. However, from previous work, we know that it is possible for there to be multiple values of k satisfying the above equation. So we define s $\geq$ k to be the largest exponent such that

$$S(p^s) = (j-1)p$$

or put another way, the product

$$1*2*3* \ldots *(j-1)p$$

contains s instances of the prime p but not s+1.

Extending the product out to

$$1*2*3*4* \ldots *(j-1)p*((j-1)p+1)*((j-1)p+2)* \ldots *jp$$

and applying lemma 1, it follows that jp is the smallest number such that (jp)! contains more than s instances of the prime p.

And so by the definition of the Smarandache function

$$S(p^{s+1}) = jp$$

Contradicting the assumption. Therefore, no such number exists and the equation

$$S(p^{k)} = np$$

is solvable in the integers.

Given that $S(p^n) = kp$ for p a prime, the next step is to determine an algorithm to effectively compute the value of the function if the input is the power of a prime. To start this, go back to the list of powers of three and realize that an additional instance of the prime is added whenever the multiplier of the prime is divisible by the prime. And if the multiplier has several instances of the factor, then each multiple will add an instance of the prime.

15

For example, consider the factorial

$$1 * \ldots * p * \ldots * 2p * \ldots * kp * \ldots * (p*2p-1)p$$

and let k represent the number of instances of p in the product. Now add the additional factor

$$(p*2p)*p$$

to this product. The addition will add 3 instances of the prime p to the product, two from the index and one since it is a multiple of p. Therefore, to compute k, the number of instances of the prime p in the product, we must include not only the contribution from the multiples of p, but also the number of times that p appears in the indices. This leads to the following lemma.

Lemma 2: Let m be an integer and consider the product (mp)!. Then the number of instances of the prime p in the product is m plus the number of times p is a factor of an index in the list of numbers

$$1, \ldots, p, \ldots, 2p, \ldots, 3p, \ldots, mp$$

The second number is computed by executing the following simple algorithm, where the division is integer division.

Step 1: Sum = 0.
Step 2: Hold = m/p.
Step 3: If hold=0 then exit with sum the desired result.
Step 4: sum=sum+hold.
Step 5: m=hold
Step 6: Go to step 2.

Proof: Clearly, p appears as a factor in the list

$$1, \ldots, p, \ldots, 2p, \ldots, mp$$

m times as a result of m products of p and an index. Performing the integer division m/p, will return the number of times that the index was of the form

$$kp$$

If we then take the integer division again (m/p)/p, we have computed the number of times that the index was of the form

16

k*p*p

And in general, performing the integer division j times will return the number of times that the index was of the form kp^j. The sum of all of the values of the integer divisions until the result is zero will yield the total contribution of instances of the prime p from the indices. Therefore, the total number of instances of the prime p in the factorial is the sum of the two numbers.

This leads us to an algorithm that will allow us to compute the value of the Smarandache function for all numbers of the form p^k, where p is prime.

Algorithm 1:

Inputs: A prime number p and an exponent k.
Output: $S(p^k)$

Step 1: If $k < p$ then exit with $S(p^k) = kp$.

Step 2: Choose an initial number startnumber, such that startnumber*p is less than or equal to the value of $S(p^k)$

Step 3: Use the algorithm in lemma 2 to determine the number of instances of the prime p in startnumber*p.

Step 4: If this number is greater than or equal to k, exit with $S(p^k)$ = startnumber*p.

Step 5: Increment startnumber by 1.

Step 6: Got to step 3.

NOTE 3: Step 1 in the above algorithm is just the default case when there are no contributions of the prime p from the indices. It is included for purposes of efficiency.

Theorem 11: $S(p^m)$ = mp if $m \leq p$. $S(p^m) <$ mp if $m > p$.

Proof: Direct consequence of the algorithm.

Theorem 12: $S(p^m) / p^m > S(p^{m+1}) / p^{m+1}$ for p a prime. Moreover,

$$\lim_{m \to \infty} \frac{S(p^m)}{p^m} = 0$$

Proof:

$$\frac{S(p^m)}{p^m} = \frac{kp}{p^m} = \frac{k}{p^{m-1}} , \quad \frac{S(p^{m+1})}{p^{m+1}} = \frac{jp}{p^{m+1}} = \frac{j}{p^m} ,$$

where $j \leq k + 1$. Therefore, the desired inequality is clear. Since the numerator increases by at most 1 when the exponent is increased by 1 and the denominator increases by a factor of at least 2, the denominator dominates the numerator rather quickly.

The final step is to determine an algorithm that will compute the value of the Smarandache function in general, and that is the topic of the next theorem.

Theorem 13: If $m = p_1^{\alpha 1} \ p_2^{\alpha 2} \ p_3^{\alpha 3} \ \ldots \ p_n^{\alpha n}$ is the prime factorization of m, then

$$S(m) = \max\{ \ S(p_i^{\alpha i}) \ \}$$

Proof: Reindex the primes so that $S(p_n^{\alpha n}) = k$ is the largest of all the function values $S(p_i^{\alpha i})$. From the properties already proven, k is then the smallest number containing the necessary number of instances of the prime p_n so that $p_n^{\alpha n}$ divides k. Since k is greater than or equal to all of the other values of $S(p_i^{\alpha i})$, it follows that k must also contain the required number of instances of the other primes so that $p_i^{\alpha i}$ divides k. Therefore, k satisfies the definition of the Smarandache function and $S(m) = k$.

Definition 4: Given $m = p_1^{\alpha 1} \ldots p_k^{\alpha k}$, we say that p_i is the **prime of concern** if

$$S(m) = S(p_i^{\alpha i})$$

Now that we have all the necessary machinery to compute the values of the Smarandache function, we can prove a simple, yet important theorem concerning the fixed points of S.

Theorem 14: $M = 4$ is the only composite number where $S(m) = m$.

Proof: Case 1: m has only one prime factor

By theorem 9, $S(p^n) = kp$. So, if n is to be a fixed point of S, then $S(p^m) = kp \Rightarrow p^m = kp$.

Subcase 1: $1 < m \leq p$

Then, by theorem 11, the equation reduces to $p^{m-1} = m$, which has only one solution, $p = 2$, $m = 2$. Note that $p^{m-1} > m$ for all other values.

Subcase 2: $m > p$.

18

Again, by theorem 11, $p^{m-1} = k$, $k < m$. And since $p^{m-1} > m$ where $p \neq 2$ or $m \neq 2$, there are no solutions.

Case 2: N has more than one prime factor.

We will deal with the case where there are two prime factors, the rest will be an obvious consequence.

Suppose $n = p_1^{\alpha 1} p_2^{\alpha 2}$ where p_1 is the prime of concern. Then $S(n) = kp_1$. If n is to be a fixed point, it follows that

$$p_1^{\alpha 1} p_2^{\alpha 2} = kp_1$$

or upon reduction

$$p_1^{\alpha 1-1} p_2^{\alpha 2} = k$$

With $k \leq m$, we then have

$$n \geq p_1^k p_2^{\alpha 2} \quad \text{where k has the form above}$$

Given the form of n, this is impossible.

If n has more than two prime factors, then the proof is similar. The only difference is that the representations have more factors.

Using theorem 13, it is now possible to construct a complete algorithm to compute the values of the Smarandache function for all numbers in the domain.

Algorithm 2:

Input: n an integer ≥ 1

Output: $S(n) = m$.

Step 1: If $n = 1$ exit with $S(n) = 0$.

Step 2: Perform the prime factorization of n.

Step 3: Use algorithm 1 to compute $S(p_i^{\alpha i})$ for all of the prime factors of n.

Step 4: Exit with $S(n)$ the maximum of those numbers computed in step 3.

19

It is a simple matter to construct a computer program to execute algorithm 2 and a C program that implements the algorithm follows. Unsigned long integers are used to store the numbers, which gives an upper limit for num in excess of 4,200,000,000. The program was compiled as a C++ program using the Borland Turbo C++ package.

```c
#include<stdio.h>
#include<math.h>

void main()
{
  long num; // The program will compute S(num)
  long temp1; // Used as a temporary storage location.
  short i,factcount; // i is used as an index variable in looping and factcount is the number
                     // of distinct prime factors of num.
  long factors[30]; // This array stores the prime factors of num.
  short exps[30]; // This array stores the exponents of the primes held in the array of
                  // factors. For example, factors[3] = 5 and exps[3] = 2 means that
                  // 5 appears twice as a factor of num.
  long smars[30]; // This array stores the values of the Smarandache function for each
                  // of the prime factors.
  long max; // This holds the maximum value of the number in the array smars[].
  long divisor; // Used as a divisor to determine prime factors of num.
  long maxcheck; // If divisor > maxcheck, then the number is prime.
  unsigned char found; // Used as a flag to exit a while loop.
  long startnum; // This variable is the one that will be repeatedly divided by the prime to
                 // determine the number of instances of the prime that were contributed
                 // by the indices.
  long sum_of_factor; // Keeps the running total of the number of instances of the prime.
  long tempsum; // Used to store the results of the repeated integer divisions.

  printf("Enter the value of the number\n");
  scanf("%ld",&num);
  if(num==1)
  {
    printf(" 1  1\n");
  }
  else
  {
    factcount=0;
    for(i=0;i<30;i++)
    {
      factors[i]=0;
      smars[i]=0;
```

20

```
        exps[i]=0;
    }

/* The first step is to break num into the equivalent prime factorization. For efficiency
   considerations we first remove all instances of the even prime 2. */

    temp1=num;
    if(num%2)
    {
        temp1=num/2;
        factors[0]=2;
        exps[0]=1;
        while(((temp1%2)==0)&&(temp1>1))
        {
            temp1=temp1/2;
            exps[0]++;
        }
        factcount=1;
    }

    maxcheck=long(sqrt(num));
    divisor=3;
    while((temp1>1)&&(divisor<=maxcheck))
    {
        if((temp1%divisor)==0)
        {
            factors[factcount]=divisor;
            exps[factcount]=1;
            temp1=temp1/divisor;
            while(((temp1%divisor)==0)&&(temp1>1))
            {
                temp1=temp1/divisor;
                exps[factcount]++;
            }
            factcount++;
        }
        divisor=divisor+2;
    }

// If temp1>1 at this point, then num is prime.
    if(temp1>1)
    {
        factcount=1;
```

```
    factors[factcount]=num;
    exps[factcount]=1;
    }

/* The next step is to compute the value of the Smarandache function for each pair of
    entries in the arrays. */

    for(i=0;i<factcount;i++)
    {
    if(exps[i]<factors[i])
    {
    smars[i]=exps[i]*factors[i];
    }
    else
    {
    startnum=exps[i]/factors[i];
    if(startnum<1)
    {
    startnum=1;
    }
    found=0;
    while(found==0)
    {
    sum_of_factor=startnum;
    tempsum=startnum/factors[i];
    while(tempsum>0)
    {
    sum_of_factor=sum_of_factor+tempsum;
    tempsum=tempsum/factors[i];
    }
    if(sum_of_factor>=exps[i])
    {
    found=1;
    }
    else
    {
    startnum++;
    }
    }
    smars[i]=startnum*factors[i];
    }

/* The final step is to determine the largest of the values of the Smarandache function. */
```

```
max=0;
for(i=0;i<factcount;i++)
{
  if(smars[i]>max)
  {
    max=smars[i];
  }
}

printf("%ld %ld\n",num,max);
 }
}
```

While most of this algorithm should be clear, there is one additional point that must be raised. For the algorithm to be efficient, the value of startnum should be chosen as close to the true value as possible. The value used here is the exponent divided by the prime. It should be obvious that the sum of all the integer divisions must be greater than or equal to this value.

Combining the results of theorems 10 and 13, we can now determine the range of the Smarandache function.

Theorem 15: Range(S) = { 0,2,3,4,5 . . . }. In other words, 1 is the only nonnegative number where there is no corresponding m such that

$$S(m) = 1$$

Proof: $S(1) = 0$. Every number larger than 1 contains a prime factor, so by theorems 10 and 13,

$$S(m) = kp \qquad \text{for p some prime divisor of n}$$

And it is not possible for 1 to be in the range of S.

Now, choose an arbitrary $m > 1$. Decompose m into the corresponding unique prime factorization and order those primes. Let q represent the smallest prime in this list and rewrite m in the form

$$m = n_1 q$$

By theorem 10 it is possible to find a number k such that

$$S(q^k) = n_1 q$$

therefore m is in the range of S. Since $m > 1$ was arbitrary, all numbers greater than 1 are in the range of S.

The primary conclusion from this theorem is that one can choose an integer n satisfying any condition(s) and as long as that integer is greater than one or zero, it is possible to find another integer m such that

$$S(m) = n$$

Therefore, problems concerning the values m such that $S(m) = n$ where n is an element of a particular set do not deal with the existence of a solution. Construction of a solution, number, and the form of solutions are the points of emphasis.

Examples:
The set of Cullen numbers is defined in the following way

$$\{ m \mid m = n*2^n + 1, \ n \geq 0 \}$$

And so, from the previous theorem, it is possible to find an integer m such that

$$S(m) = n*2^n + 1 \quad \text{for } n > 0$$

The set of Catalan numbers is defined in the following way

$$c_1 = 1, \quad c_n = \frac{1}{n} \binom{2n-2}{n-1} \quad \text{for } n \geq 2$$

again, from the previous theorem, there is a number m such that

$$S(m) = c_k$$

for any Catalan number c_k not equal to 1.

If p is any prime, we already know that $NS(p)$ is at least

$$2^{\pi(p)-1}$$

With the results of theorem 13, we can now continue to the examine the size of $NS(n)$.

If $S(m) = S(p_i^k) = np_i$ where p_i is the prime of concern, we know that if p is any prime less than p_i , then $S(pm) = S(m)$.

Therefore, we can now extend this count to include

$$NS(n) \text{ is at least } 2^{\pi(n)}$$

for n composite.

Furthermore, if $q < p$ where q is also prime and $S(q^k) < p$, then

$$S(pq^k) = S(p) = p$$

so the value mentioned above for NS(n) is less than the true size.

Before, we continue, a well-known result concerning the growth in the size of sets must be mentioned.

Counting Principle 2: Let L be a set containing k elements. Then the number of distinct subsets that can be formed from the elements of L is

$$2^k$$

and if an additional element is added that number grows to

$$2^{k+1}$$

Of course, elimination of an element reduces the number to

$$2^{k-1}$$

This principle is well-known, so no proof need be given here. Those interested in a proof should consult a book on combinatorics.

Let p be an odd prime, which forces p - 1 to be even. Therefore, p! would contain at least

$$(p - 1) / 2$$

instances of 2 as a factor. From this, it follows that

$$2^{(p-1)/2} \text{ divides p!}$$

by theorem 9 $S(2^{(p-1)/2}) \leq \frac{p-1}{2} * 2 = p - 1$

And by theorem 13,

$$S(2^k p) = S(p) = p$$

for k $\leq$ $\frac{p-1}{2}$.

If we remove the prime number two from consideration, then the number of possible combinations of all primes less than p is

$$2^{\pi(p)-2}$$

Combining 2 and then 2^2 to this set of primes leads to the inclusion of 2^3, so the addition of each power of two does not double the number of possible combinations. In this case we can only choose one item from the set of powers of two at a time. Therefore, to compute the number of possible combinations, we need another fundamental principle of counting.

Counting Principle 3: If we have a set of objects of size m and another set of objects of size n and we can only choose one object from the second set at a time, then the number of possible ways that this can be done is given by

$$m * n$$

This principle is well-known, so no proof is given here. Again, those interested in a proof should consult a book on combinatorics.

We can now prove the following expansion of the computation of NS(p).

Theorem 16: If p is an odd prime, then the number of distinct solutions m to the equation

$$S(m) = p$$

is at least

$$2^{\pi(p)-2} * \frac{p-1}{2}$$

Proof: Apply counting principle 2 to the numbers computed above.

Example: For p = 1009, the first 4 digit prime, there are at least

$$2^{87} * \frac{1008}{2} = 2^{87} * 504$$

numbers m such that S(m) = 1009.

However, this also understates the number of solutions, in that it includes only the instances of 2 that are contributed by the second numbers in the products

$$2k$$

completely ignoring the contributions when 2 divides the index k. This is significant for even small numbers, as

$$S(2^{64}) = 66$$

and not 128, as is used in theorem 16.

There is one last item to deal with before we move on to the examination and clarification of some of the problems involving the Smarandache function. While unsigned long integers can accomodate numbers in excess of 4,000,000,000, there are times when it is necessary to examine particular numbers larger than this. For example, one may be interested in the behavior of the set

{ S(m) | where m is a Cullen number }

And the C++ program may not allow for adequate evaluation.

UBASIC is a public domain extended precision package that allows numbers to have thousands of digits. As the name implies it is a subset of BASIC, so it is very easy to learn.

The following program is a UBASIC implimentation of the algorithm to compute the values of the Smarandache function. Num is the input number and is hardcoded into the program.

```
10 dim facts(30),eps(30),smar(30)
50 num=3
100 for i=1 to 30
110 facts(i)=0
120 eps(i)=0
130 smar(i)=0
140 next i
150 factcount=1
160 j=2
170 t1=num
175 ' \ is integer division in UBASIC
180 t2=t1\j
185 'The remainder of integer division is automatically stored in the identifier res
190 if res<>0 then goto 400
200 facts(factcount)=j
210 eps(factcount)=1
220 t1=t2
```

27

```
230 if t1=1 then goto 500
240 t2=t1\j
250 if res<>0 then goto 300
260 eps(factcount)=eps(factcount)+1
270 goto 220
300 factcount=factcount+1
400 j=j+1
410 goto 180
500 print"The number is ";num
510 for i=1 to factcount
520 print facts(i),eps(i)
530 next i
690 smar=1
700 for i=1 to factcount
710 p=facts(i)
720 a=eps(i)
730 gosub 20000
740 if spa<smar then goto 760
750 smar=spa
760 next i
770 print"The value of the function is ";smar
2000 end
20000 'Subroutine to compute s(p^a)
20005'The value of the function is spa
20010 if a>=p then goto 20040
20020 spa=p*a
20030 return
20040 n1=a\2
20050 s1=n1
20055 t1=n1
20060 t2=t1\p
20070 if res<p then goto 20105
20080 s1=s1+t2
20090 t1=t2
20100 goto 20060
20105 s1=s1+t2
20100 goto 20060
20110 if s1>=a then goto 20140
20120 n1=n1+1
20130 goto 20050
20140 spa=p*n1
20150 return
```

Those familiar with old, original BASIC should have little trouble understanding this program.

At this point, we have completed the introduction of the function and how to compute the values. Although the Smarandache function is of relatively recent vintage, a large number of problems have already been posed concerning the properties and consequences of the function. With the information that we now have at our disposal, we can proceed to examine and clarify some of these problems. The computer programs listed above were also used to search for solutions.

Chapter 2

In this chapter, we introduce and resolve several questions that have been raised concerning the Smarandache function. Most of the following problems have appeared in either [2] or [3].

Theorem 17: For any integer n $\geq$ 0, it is possible to find another integer m such that

$$S(m) = n!$$

Proof: $S(1) = 0! = 1!$ $S(2) = 3$ $S(3) = 3$

Let p $\leq$ n be a prime. Then, since n! = kp for k some integer, we apply theorem 10 and conclude that there is some value m, such that

$$S(p^m) = kp = n!$$

One thing that is often done with functions is to perform repeated iterations, such as f(f(f(x))). It is not difficult to analyze what happens when this is done with the Smarandache function.

Definition 5: Let $S^k(n)$ be used to represent k iterations of the function S.

$$S(S(\ldots S(n) \ldots))$$

Definition 6: A number n is said to be a fixed point of the function f if

$$f(n) = n$$

Lemma 3: If n = 4 or n prime, then $S^k(n) = n$ for any number k.

Proof: By theorem 3 S(p) = p and by theorem 14 n = 4 is the only composite fixed point.

NOTE 4: From this, S^k has an infinite number of fixed points.

To understand what happens when p is not a prime, we need the following two general results.

Lemma 4: If n is has at least two distinct prime factors, then S(n) $\leq$ n/2.

Proof: By hypothesis, n can be factored into two or more prime components. By theorem 13, S(n) is the maximum value of the separate $S(p_i^k)$ values. Since the factors other than p_i

where p_i is the prime of concern, must have a product greater than or equal to 2, the following holds.

$$p_i^k \leq n/2$$

Applying theorem 1, the inequality becomes

$$S(p_i^k) \leq p_i^k \leq n/2$$

Lemma 5: If $n = p^k$, where $k > 1$ and $n > 4$ then $S(n) < n$.

Proof: There are two cases.

Subcase 1: $k \leq p$

By theorem 11, $S(n) = kp$. Then since $k \leq p^{k-1}$ for $k > 1$ and $p \geq 2$ we have $kp \leq p^k$. Equality occurs only when $k = 2$ and $p = 2$.

Subcase 2: $k > p$

Applying theorem 11, $S(n) = mp < kp$. Reasoning similar to that of subcase 1 allows us to conclude that $kp < p^k$ so it follows that $mp < p^k$.

Combining the results of the previous three lemmas we have the following general behavior when the Smarandache function is iterated.

Theorem 18:

a) If $m = 1$, then $S^k(m)$ is undefined for $k > 1$.

b) If $m > 1$, then $S^k(m) = n$ where n is 4 or prime for all k sufficiently large.

Proof:
(a) Obvious, since $S(1) = 0$ and 0 is not in the domain of S.

(b) If $m = p$ a prime, apply lemma 3. If $m = 4$, use theorem 14.

Therefore, if m is not prime or 4, we can apply lemmas 4 and 5 to conclude that the result of an iteration is strictly less than the input number. Repeated iteration of successive composite numbers will then move towards zero. Obviously this cannot continue indefinitely.
Since the result of an iteration is of the form kp, where p is the prime of concern, the final

value must be a prime to the first power or 4.

Another interesting consequence of the repeated iteration is that while every process starting at a number $m > 1$ terminates at a fixed point, there is no maximum number of steps that it can take.

Theorem 19: There is no number K such that for every number $m > 1$,

$$S^K(m) = n \text{ where n is a fixed point of S}$$

Proof: Suppose there is some number K that satisfies the conditions of the theorem. Then, there must exist some non-empty set of numbers where S must be iterated K times to reach a fixed point. Let m be an arbitrary element of that set. Clearly, $m > 1$. From theorem 15, we know that there is another number r such that

$$S(r) = m$$

Therefore, it must take $K + 1$ iterations of S from the initial number r to reach a fixed point. This contradiction forces the conclusion that no such K exists.

We already know that there are an infinite number of fixed points of S. The next question concerns how many numbers will iterate to a given fixed point.

Theorem 20: Let $m > 2$ be a fixed point of S. Then the set of numbers

$$U = \{ n \mid S^k(n) = m \}$$

is infinite.

Proof: As the "oddest" of the fixed points, we deal with 4 separately.

By computation $S(8) = 4$ and $S(64) = 8$. Assume that U is finite and that $n = 2^k$ $k > 2$ is the largest power of two found in U. From theorem 10 we know that there is some number 2^j such that

$$S(2^j) = 2^k$$

Since 2^j is composite, and $j > 2$, it follows that $j > k$. This contradiction forces the conclusion that U is infinite.

Let $p > 2$ be a prime. Clearly, $S(p^2) = 2p$ and $S(2p) = p$ by choice of p. Being a multiple of 2, theorem 10 allows us to conclude that there is some number $k > 2$ such that

32

$$S(2^k) = 2p$$

And we can use reasoning similar to that for the previous case to conclude that there is no largest power of two that iterates to p.

Corollary 4: N = 2 is the only number where

$$S^k(n) = 2$$

Proof: If p is prime, then the only numbers m such that S(m) = p are m = p or those products kp, where $S(k) \leq p$. If k contains an odd prime, S(kp) = 2 is impossible. Therefore, the only possibility is if k is a power of 2. However, any number of that form must have more than one instance of 2 and iterate to either another prime or the fixed point 4.

The modifcations of the number theoretic functions $\Phi(n) \pm 1$ and $\sigma(n) \pm 1$ have also been iterated. Some unsolved problems stemming from these operations are given in the book by Guy[4]. Our next topic will be the behavior of S(m) - 1 as it is iterated and we will start with an important lemma.

Lemma 6: S(m) - 1 iterated twice is less than m if m > 5.

Proof: There are two cases

   Case 1: m = p a prime. Then S(p) = p, which forces p - 1 to be composite. Applying
   lemmas 4 and 5 S(p-1) $<$ p-1 and we can construct the desired inequality.

   Case 2: m is composite. Then by lemmas 4 and 5, S(m) $<$ m and S(m) - 1 $<$ m.
   Applying theorem 1, we then have

$$S(S(m) - 1) - 1 \; < \; m$$

From this, it is clear that repeated iteration of S(m) - 1 will lead to a continuing reduction in size. Since the repeated iteration must lead to a value that is eventually $\leq$ 5, let us now examine what happens in that range.

S(5) - 1 = 4
S(4) - 1 = 3
S(3) - 1 = 2
S(2) - 1 = 1
S(1) - 1 = -1

Leading to the theorem for iterations of S(m) - 1.

Theorem 21: Repeated iteration of the function S(m) -1 always leads to a terminal value of -1, where the operation is no longer defined.

Proof: Clear from previous work.

For the function S(m) + 1 we compute the first few values

$$
\begin{aligned}
S(1) + 1 &= 1 \\
S(2) + 1 &= 3 \\
S(3) + 1 &= 4 \\
S(4) + 1 &= 5 \\
S(5) + 1 &= 6 \\
S(6) + 1 &= 4 \\
S(7) + 1 &= 8 \\
S(8) + 1 &= 5 \\
S(9) + 1 &= 7 \\
S(10) + 1 &= 6
\end{aligned}
$$

and it is clear that the iteration of S(m) + 1 will enter a loop if the value ever drops to 10 or less.

To determine what occurs when S(m) + 1 is iterated twice, we need the following two lemmas.

Lemma 7: If m is composite and $m > 10$, then $S(m) < m - 2$.

Proof: If m contains more than one prime factor, then we can apply lemma 4. If $m = p^k$ we need only apply the reasoning of lemma 5 with the understanding that $k > 3$ or $p > 2$.

Lemma 8: S(m) + 1 iterated twice is less than m for $m > 10$.

Proof: We split the proof into cases.

Case 1: m = p where p is prime. Then $S(p) + 1 = p+1$ where p must be composite. By lemma 7, $S(p+1) < p - 1$, so $S(p+1) + 1 < p - 1 + 1 = p$.

Case 2: m is composite. If m contains more than one prime factor, by lemma 4

$$S(m) \leq m/2 \Leftrightarrow S(m) + 1 \leq m/2 + 1$$

$$S(m/2+1) + 1 \quad \text{is then at most} \quad m/2 + 2$$

34

which by choice of m is less than m. If m contains only one prime factor then by lemma 7

$$S(m) + 1 \ < \ m - 2$$

and $S(S(m)+1) \ < \ m - 1$

All of which leads to the following

Theorem 22: Repeated iteration of the function $S(m) + 1$ leads to a loop if the initial number is greater than 1.

Proof: By lemmas 7 and 8, if $m \ > \ 10$ each iteration reduces the value. At some point, the value must drop under 10 and the repeated behavior can be seen from the computations for those numbers.

It is not difficult to see that the behavior of the iteration of $S(m) - 1$ can be generalized.

Theorem 23: Iteration of $S(m) - k$ for $k \geq 1$ will always lead to smaller numbers with eventual termination at a number not in the domain of S.

Proof: The inequalities in the proof of $S(m) - 1$ also hold for any such k in the place of 1.

Another useful result concerns the behavior of the differences of successive values of the function.

Theorem 24: The set of numbers $\{ \ d \mid d = \mid S(n+1) - S(n) \mid \ \}$ is unbounded.

Proof: Suppose there is some number M such that

$$M = \max \{ \ \mid S(n+1) - S(n) \mid \ \}$$

Since the number of primes is infinite, we can find some prime p such that $p > 4M$. By theorem 3, $S(p) = p$. Since p-1 is composite, we can apply lemma 4 to conclude that $S(p-1) \leq (p-1)/2$. Therefore,

$$S(p) - S(p-1) \geq p - (p-1)/2 = (p+1)/2 > M$$

contradicting the choice of M.

The following is then a direct consequence of this theorem.

Theorem 25: S does not satisfy the Lipschitz condition, i.e. there is no number $M > 0$

such that $|S(m) - S(n)| \leq M|m - n|$.

Proof: With the understanding that $|m - n| = 1$ in the previous theorem, this is a direct consequence of that theorem.

Corollary 5: S does not satisfy a k-Lipschitz condition. I.e.

$$|S(m) - S(n)| \leq M(|m - n|)^k , \quad \text{for k some integer.}$$

Proof: Direct consequence of the fact that $|m - n|^k = 1$ in the previous two theorems.

Now that we know that the differences $|S(n+1) - S(n)|$ have no upper bound, the next step is to determine what the lower bound is. Since $S(2) = 2$ and $S(3) = 3$, we know that it is at most 1. The next question, which is our first unsolved problem, is if the difference can ever be zero.

Unsolved Problem 1: Is there a number n such that $S(m) = S(m+1)$?

It has been conjectured in [5] that there is no such number.

The author used a modification of the C program given previously and found no solutions up through num $\leq 1,000,000$. As we will see shortly, that fact is essentially meaningless when dealing with the values of S.

By lemma 4, lemma 5 and theorem 3, it is clear that if n is such a number, then neither n or n+1 can be prime. Also, the numbers cannot share any common prime factors. Since both numbers must be composite, $S(n) = kp$, where p is the prime of concern for n and $S(n+1) = rq$, where q is the prime of concern for n+1. For the equality

$$kp = rq$$

to hold, it follows that q would have to divide k and p divide r. That it may be possible to find a solution by severely restricting the search is easily justified. Consider an arbitrary number

$$2*3*n$$

where n $\geq$ 1. By theorem 10 it is possible to find numbers k1 and k2 such that

$$S(2^{k1}) = 2*3*n = S(3^{k2})$$

Therefore, there are an infinite number of simultaneous powers of 3 and 5 such that their respective values of S are equal. It also follows that if p is any prime $5 < p < 3*5*n$, then

$$S(p*2^{k1}) = 2*3*n = S(p*3^{k2})$$

And in general, any number m ,where $S(m) \leq 2*3*n$ can be placed in the product without affecting the value of S.

Which gives us the following algorithm to search for solutions to the equation

$$S(m) = S(m+1)$$

1. Start with any number evenly divisible by 2 and 3. Call that value num.

2. Determine the values k1 and k2 such that $S(2^{k1}) = num = S(3^{k2})$.

3. Construct the set P of all primes $p < num$.

4. Construct the set Q of all distinct products of elements of P.

5. Construct all products $r*2^{k1}$ and $s*3^{k2}$ where r and s are elements of Q. If the difference of these products is ever 1, exit with the products as the solution.

6. Increment num by $2*3$

7. Go to step 2.

Although this algorithm was constructed using the example primes 2 and 3, it will also work if we use p and q, two arbitrary primes. However, we must be careful to maintain the proper parity, which can be used to improve the performance. Since 2^{k1} is even and 3^{k2} is odd the products constructed in step 5 are all of opposite parity. Which is what is needed since n and n+1 are of opposite parity.

However, if we use two odd primes, then the powers are both of odd parity and the products r and s must be of opposite parity. This is easily repaired as 2 is available to build the products.

As we have already seen, the number of all possible products of primes rises very rapidly as num increases. Furthermore, while not included in the algorithm, it would be possible to include all powers p^k where $S(p^k) \leq num$.

Given the large number of possible combinations that develop as the numbers get larger, the author is convinced that solutions to the equation do exist and makes the following conjecture.

Conjecture 1: There are numbers n and n+1 such that

$$S(n) = S(n+1)$$

It is the authors fervent hope that readers will search for solutions and one will be found. However, there will also be no disappointment if a proof that there are no solutions is discovered.

Additional families of unsolved problems concerning the algebraic relationships between successive values of the Smarandache function have been posed. Very little has actually been resolved concerning these issues, although at times there are tantalizing hints as to a possible resolution.

Definition 7: The Fibonacci sequence is defined in the following way:

$$F(0) = 0 \quad F(1) = 1 \quad F(n+2) = F(n+1) + F(n) \quad \text{for } n > 1.$$

In 1994, T. Yau[6] noted that $S(9) + S(10) = S(11) \Leftrightarrow 6 + 5 = 11$, $S(119) + S(120) = S(121) \Leftrightarrow 17 + 5 = 22$ and asked if there were any other solutions matching the Fibonacci-like behavior.

The author conducted an additional computer search up through n $\leq$ 1,000,000 and found the additional solutions

$S(4900) + S(4901) = S(4902) \Leftrightarrow$
$S(2*2*5*5*7*7) + S(13*13*29) = S(2*3*19*43) \Leftrightarrow 14 + 29 = 43$

$S(26243) + S(26244) = S(26245) \Leftrightarrow$
$S(7*23*163) + S(2*2*3*3*3*3*3*3*3) = S(5*29*181) \Leftrightarrow 163 + 18 = 181$

$S(32110) + S(32111) = S(32112) \Leftrightarrow$
$S(2*5*13*13*19) + S(163*197) = S(2*2*2*2*3*3*223) \Leftrightarrow$
$26 + 197 = 223$

$S(64008) + S(64009) = S(64010) \Leftrightarrow$
$S(2*2*2*3*3*7*127) + S(11*11*23*23) = S(2*5*37*173) \Leftrightarrow$
$127 + 46 = 173$

$S(368138) + S(368139) = S(368140) \Leftrightarrow$
$S(2*23*53*151) + S(3*41*41*73) = S(2*2*5*79*233) \Leftrightarrow$
$151 + 82 = 233$

$S(415662) + S(415663) = S(415664) \Leftrightarrow$

$S(2*3*13*73*73) + S(19*131*167) = S(2*2*2*2*83*313) \Leftrightarrow$
$146 + 167 = 313$

In examining the solutions, notice that with only two exceptions, the values of $S(n)$ are determined by either the largest prime factor or from a prime factor that occurs twice. This certainly makes sense, as we cannot have

odd prime + odd prime = odd prime

In nearly all other cases, we have the prime of concern appearing only to the first power.

To begin our analysis, we note that all of the following

odd prime + odd prime = twice an odd prime

twice an odd prime + odd prime = odd prime

odd prime + twice an odd prime = odd prime

can serve as the foundation. Since we know that $S(p) = p$ and $S(p^2) = 2p$ for p prime, finding solutions where

$$S(r) + S(s) = S(t)$$

is just a matter of finding primes p_1, p_2 and p_3 such that

$$p_1 + p_2 = 2p_3 \quad \text{or}$$

$$2p_1 + p_2 = p_3 \quad \text{or}$$

$$p_1 + 2p_2 = p_3$$

Starting with primes that satisfy any of the three formulas above and then taking the list of all primes such that $S(p1) < r$, $S(p2) < s$ and $S(p3) < t$, it is possible to create many triples (r,s,t) such that

$$S(r) + S(s) = S(t)$$

And as has already been noted, the size of each list is bounded below by a power of two. Since the lists are mutually exclusive, if we exclude everything but the primes, the number of possible combinations is at least

39

$$2^k \quad \text{where} \quad k = \pi(p_1) + \pi(p_2) + \pi(p_3)$$

Inclusion of the powers of the primes will substantially increase the number of possibilities. Note that the parity of the the numbers (r,s,t) has no affect on whether the formula is satisfied.

The final parameter to be satisfied is then to find a combination

$$(r,s,t) = (n, n+1, n+2)$$

Since 2 is the number with the largest allowable number of repeats, this could explain the number of solutions with several instances of 2 as a factor. Also, the rapidly growing number of possible combinations allows one to argue that the number of solutions will grow as larger numbers are used. The maximum number of 1,000,000 in the computer search is not large enough to exhaust all possible combinations using the seed primes of the solutions found. All of this leads to the conjecture.

Conjecture 2: There are an infinite number of solutions to the equation

$$S(n) + S(n+1) = S(n+2)$$

Rationale: In addition to the arguments presented above, note that if there are three primes in an arithmetic progression

$$p, \ p + d, \ p + 2d$$

then

$$S(p) + S(p + 2d) = S((p+d)^2)$$

and we have a solution satisfying the initial parameter. It is known that there are an infinite number of such sequences[7].

Not only are each of the products created by a collection of the primes less than a number distinct, but an examination of the clustering of those products is of value. Clearly, the minimum and maximum values of the products are

$$2 \quad \text{and} \quad \prod_{k=1}^{\pi(p)-1} p_k \ .$$

However, if the locations of the values are plotted, most will be concentrated somewhere in the "middle" of this range, increasing the likelihood of finding a desired triple. If we are starting with numbers that are primes but close together, the probability of constructing three numbers n, n+1 and n+2 is even higher.

Since we are on the subject of Fibonacci numbers, it is a good time to explore a related problem. Given the range of S, for any Fibonacci number F_k, where $k > 2$, it is possible to find another number m such that

$$S(m) = F_k$$

Therefore, questions concerning a relationship between S and F must have additional parameters. A logical question to explore asks for Fibonacci numbers on both sides.

Problem 1: How many pairs of Fibonacci numbers (F_j, F_k) are there such that

$$S(F_j) = F_k.$$

Clearly, if F_j is prime, then (F_j, F_j) is a solution. As is mentioned in [8], it is not and may never be known if there are an infinite number of Fibonacci numbers that are prime.

The next step is then to ask the related question with the restriction that one or both Fibonacci numbers be composite. Following previous work, if m is any number such that $S(m) \leq F_j$, then (mF_j, F_j) is also a solution. Therefore, it is possible to have solutions where the first is composite, but here again the number of solutions is dependent on the number of prime Fibonacci numbers. It should be obvious that it is not possible to have a solution where the first is prime and the second is not.

Which takes us to the final option, where both numbers are composite. We already know that for any number, sufficiently large, of the form $F_k = np$ where p is prime, there is some value of k such that

$$S(p^k) = np$$

While we have no guarantee that p^k is also a Fibonacci number we know that it is possible to find a very large number of composite numbers m such that $S(m) = np$. This is used to justify the following conjecture.

Conjecture 3: There are an infinite number of pairs of Fibonacci numbers (F_i, F_k) such that

$$S(F_i) = F_k.$$

The Lucas numbers are defined in a very similar way

$L_0 = 2 \quad L_1 = 1$

$L_{n+2} = L_{n+1} + L_n$

and the same conjecture can be made concerning this sequence.

Another group of problems concerns the additivity of successive values of the Smarandache function and was first brought to the author's attention in a pre-publication manuscript by Jorge Castillo[9] that was passed on by R. Muller.

We start with the first, which is related to the Fibonacci question addressed earlier.

Problem 2: How many quadruplets satisfy the relationship

$$S(n) + S(n+1) = S(n+2) + S(n+3) \ ?$$

And the three solutions

$S(6) + S(7) = S(8) + S(9) \quad \Leftrightarrow \quad 3 + 7 = 4 + 6$
$S(7) + S(8) = S(9) + S(10) \quad \Leftrightarrow \quad 7 + 4 = 6 + 5$
$S(28) + S(29) = S(30) + S(31) \quad \Leftrightarrow \quad 7 + 29 = 5 + 31$

were listed.

The aforementioned C program was again put to work searching for additional solutions. Due to the large number, the run was terminated at 100,000. The complete list of new solutions is given below

$S(114) + S(115) = S(116) + S(117) \quad \Leftrightarrow \quad 19 + 23 = 29 + 13 = 42$
$S(1720) + S(1721) = S(1722) + S(1723) \quad \Leftrightarrow \quad 43 + 1721 = 41 + 1723$
$S(3538) + S(3539) = S(3540) + S(3541) \quad \Leftrightarrow \quad 61 + 3539 = 59 + 3541$
$S(4313) + S(4314) = S(4315) + S(4316) \quad \Leftrightarrow \quad 227 + 719 = 863 + 83 = 946$
$S(8474) + S(8475) = S(8476) + S(8477) \quad \Leftrightarrow \quad 223 + 113 = 163 + 173 = 336$
$S(10300) + S(10301) = S(10302) + S(10303) \quad \Leftrightarrow \quad 103 + 10301 = 101 + 10303 = 10404$
$S(13052) + S(13053) = S(13054) + S(13055) \quad \Leftrightarrow \quad 251 + 229 = 107 + 373 = 480$
$S(15417) + S(15418) = S(15419) + S(15420) \quad \Leftrightarrow \quad 571 + 593 = 905 + 257 = 1164$
$S(15688) + S(15689) = S(15690) + S(15691) \quad \Leftrightarrow \quad 53 + 541 = 523 + 71 = 594$
$S(19902) + S(19903) = S(19904) + S(19905) \quad \Leftrightarrow \quad 107 + 1531 = 311 + 1327 = 1638$
$S(22194) + S(22195) = S(22196) + S(22197) \quad \Leftrightarrow \quad 137 + 193 = 179 + 151 = 330$
$S(22503) + S(22504) = S(22505) + S(22506) \quad \Leftrightarrow \quad 577 + 97 = 643 + 31 = 674$
$S(24822) + S(24823) = S(24824) + S(24825) \quad \Leftrightarrow \quad 197 + 241 = 107 + 331 = 438$
$S(26413) + S(26414) = S(26415) + S(26416) \quad \Leftrightarrow \quad 433 + 281 = 587 + 127 = 714$
$S(56349) + S(56350) = S(56351) + S(56352) \quad \Leftrightarrow \quad 2087 + 23 = 1523 + 587 = 2110$

$S(70964) + S(70965) = S(70966) + S(70967) \Leftrightarrow 157 + 83 = 137 + 103 = 240$
$S(75601) + S(75602) = S(75603) + S(75604) \Leftrightarrow 173 + 367 = 79 + 461 = 540$
$S(78610) + S(78611) = S(78612) + S(78613) \Leftrightarrow 1123 + 6047 = 6551 + 619 = 7170$
$S(86505) + S(86506) = S(86507) + S(86508) \Leftrightarrow 79 + 167 = 157 + 89 = 246$

Looking at the solutions in this case, a clear pattern emerges. The typical solution consists of four numbers where the prime of concern appears once and the difference of two primes of concern matches the difference of the other two. For example,

$$19902 = 2*3*31*107 \qquad 19903 = 13*1531 \qquad 19904 = 2*2*2*2*2*2*311$$

$$19905 = 3*5*1327$$

and $\qquad 1531 - 1327 = 204 = 311 - 107$

Conjecture 4: There are an infinite number of values of n such that

$$S(n) + S(n+1) = S(n+2) + S(n+3)$$

Rationale: Choose two pairs of primes p_1, p_2, p_3 and p_4 such that $p_1 - p_2 = p_3 - p_4$. Then if we can find collections of primes C_1, C_2, C_3 and C_4 such that p_1, p_2, p_3 and p_4 remain the primes of concern in the products

$$C_1 p_1 , \ C_2 p_2, \ C_3 p_3 \text{ and } C_4 p_4$$

and $C_1 p_1$, $C_4 p_4$, $C_3 p_3$, and $C_2 p_2$ are consecutive integers, we will have a solution

The known rate of growth of $\pi(x)$ guarantees that there are an infinite number of quadruples (p_1, p_2, p_3, p_4) satisfying the initial condition. Also, the number of possible products C_i grows at a rapid rate, increasing the likelihood of a solution.

The following companion problem was also mentioned in the Castillo manuscript,

Problem 3: How many solutions are there to the relationship

$$S(n) - S(n+1) = S(n+2) - S(n+3)$$

with the three solutions

$S(1) - S(2) = S(3) - S(4) \Leftrightarrow 1 - 2 = 3 - 4$
$S(2) - S(3) = S(4) - S(5) \Leftrightarrow 2 - 3 = 4 - 5$
$S(49) - S(50) = S(51) - S(52) \Leftrightarrow 14 - 10 = 17 - 13$

Performing another computer search up through 100,000, the following additional solutions were found,

$S(40) - S(41) = S(42) - S(43) \Leftrightarrow 5 - 41 = 7 - 43 = -36$
$S(107) - S(108) = S(109) - S(110) \Leftrightarrow 107 - 9 = 109 - 11 = 98$
$S(2315) - S(2316) = S(2317) - S(2318) \Leftrightarrow 463 - 193 = 331 - 61 = 270$
$S(3913) - S(3914) = S(3915) - S(3916) \Leftrightarrow 43 - 103 = 29 - 89 = -60$
$S(4157) - S(4158) = S(4159) - S(4160) \Leftrightarrow 4157 - 11 = 4159 - 13 = 4146$
$S(4170) - S(4171) = S(4172) - S(4173) \Leftrightarrow 139 - 97 = 149 - 107 = 42$
$S(11344) - S(11345) = S(11346) - S(11347) \Leftrightarrow 709 - 2269 = 61 - 1621 = -1560$
$S(11604) - S(11605) = S(11606) - S(11607) \Leftrightarrow 967 - 211 = 829 - 73 = 756$
$S(11968) - S(11969) = S(11970) - S(11971) \Leftrightarrow 17 - 11969 = 19 - 11971 = -11952$
$S(13244) - S(13245) = S(13246) - S(13247) \Leftrightarrow 43 - 883 = 179 - 1019 = -840$
$S(15048) - S(15049) = S(15050) - S(15051) \Leftrightarrow 19 - 149 = 43 - 173 = -130$
$S(19180) - S(19181) = S(19182) - S(19183) \Leftrightarrow 137 - 19181 = 139 - 19183 = -19044$
$S(19692) - S(19693) = S(19694) - S(19685) \Leftrightarrow 547 - 419 = 229 - 101 = 128$
$S(26219) - S(26220) = S(26221) - S(26222) \Leftrightarrow 167 - 23 = 2017 - 1873 = 144$
$S(29352) - S(29353) = S(29354) - S(29355) \Leftrightarrow 1223 - 197 = 1129 - 103 = 1026$
$S(29415) - S(29416) = S(29417) - S(29418) \Leftrightarrow 53 - 3677 = 1279 - 4903 = -3624$
$S(43015) - S(43016) = S(43017) - S(43018) \Leftrightarrow 1229 - 283 = 1103 - 157 = 946$
$S(44358) - S(44359) = S(44360) - S(44361) \Leftrightarrow 7393 - 6337 = 1109 - 53 = 1056$
$S(59498) - S(59499) = S(59500) - S(59501) \Leftrightarrow 419 - 601 = 17 - 199 = -182.$

In this case, a pattern emerges that is similar to the previous problem.

$59498 = 2*71*419 \qquad 59499 = 3*3*11*601$
$59500 = 2*2*5*5*5*7*17 \qquad 59501 = 13*23*199$

where each prime of concern appears once in each number and the differences of the pairs of primes match.

Conjecture 5: There are an infinite number of solutions to the equation

$$S(n) - S(n+1) = S(n+2) - S(n+3)$$

Rationale: Similar to that used to justify conjecture 4.

Another problem mentioned in the Castillo manuscript is an extension of the additive sequence

Problem 4: How many solutions are there to the relationship

$$S(n) + S(n+1) + S(n+2) = S(n+3) + S(n+4) + S(n+5)$$

with a single solution given

$$S(5) + S(6) + S(7) = S(8) + S(9) + S(10) \Leftrightarrow 5 + 3 + 7 = 4 + 6 + 5.$$

A computer search up through 100,000 yielded the additional solutions

$$S(5182) + S(5183) + S(5184) = S(5185) + S(5186) + S(5187) \Leftrightarrow$$
$$2591 + 73 + 9 = 61 + 2593 + 19 = 2673$$

$$S(9855) + S(9856) + S(9857) = S(9858) + S(9859) + S(9860) \Leftrightarrow$$
$$73 + 11 + 9857 = 53 + 9859 + 29 = 9941$$

$$S(10428) + S(10429) + S(10430) = S(10431) + S(10432) + S(10433) \Leftrightarrow$$
$$79 + 10429 + 149 = 61 + 163 + 10433 = 10657$$

$$S(28373) + S(28374) + S(28375) = S(28376) + S(28377) + S(28378) \Leftrightarrow$$
$$1669 + 4729 + 227 = 3547 + 1051 + 2027 = 6625$$

$$S(32589) + S(32590) + S(32591) = S(32592) + S(32593) + S(32594) \Leftrightarrow$$
$$71 + 3259 + 109 = 97 + 2963 + 379 = 3439$$

$$S(83323) + S(83324) + S(83325) = S(83326) + S(83327) + S(83328) \Leftrightarrow$$
$$859 + 563 + 101 = 683 + 809 + 31 = 1523$$

While the number of solutions is much less than that for the sums of two values, a similar pattern emerges. For example,

$32589 = 3*3*3*17*71$	$32590 = 2*5*3259$
$32591 = 13*23*109$	$32592 = 2*2*2*2*3*7*97$
$32593 = 11*2963$	$32594 = 2*43*379$

where each prime of concern appears only once in each product. However, in this case, there is no common difference of two primes to start the construction. The basic construction here starts with six primes such that

$$p_1 + p_2 + p_3 = p_4 + p_5 + p_6$$

and then finding six collections of numbers C_1, C_2, C_3, C_4, C_5 and C_6 such that each of the products $C_i p_i$ has the original prime as the prime of concern, and where the products are consecutive integers. With the tighter parameters of the sum of three primes two different ways and having to find 6 collections to make the proper sums, it should come as no surprise that the number of solutions discovered is smaller than that for the earlier relationship.

The three previous examples of problems appearing in the paper by Castillo were all particular instances of a general family of problems that were also defined in the paper.

Definition 8: A relationship of the form

$$S(n) \ \nabla \ S(n{+}1) \ \nabla \ldots \nabla \ S(n{+}p) = S(n{+}p{+}1) \ \nabla \ldots \nabla \ S(n{+}p{+}q)$$

where ∇ represents a generic operation defined on the integers, is said to be a Smarandache p-q-relationship.

Example: The Fibonnaci-like expression $S(n) + S(n{+}1) = S(n{+}2)$ would be a Smarandache 2-1 relationship.

Definition 9: A solution to a Smarandache p-q relationship for a specific operator is said to be a Smarandache p-q-{name of operator} sequence.

Example: $S(83323) + S(83324) + S(83325) = S(83326) + S(83327) + S(83328) \Leftrightarrow$
$$859 + 563 + 101 = 683 + 809 + 31 = 1523$$

would be an example of a Smarandache 3-3-additive relationship.

Which brings us to the first of several families of problems of this type that can be posed.

Problem 5: For what values of k is there a solution to the additive relationship

$$S(n) + S(n{+}1) + \ldots + S(n{+}k) = S(n{+}k{+}1) \ ?$$

The case of $k = 1$ is the Fibonacci-like sequence and has already been resolved in the affirmative.

A computer search for solutions to the $k = 2$ relationship, sometimes called the Tribonnaci sequence

$$S(n) + S(n{+}1) + S(n{+}2) = S(n{+}3)$$

up through 100,000 yielded the following list.

$S(20) + S(21) + S(22) = S(23) \Leftrightarrow 5 + 7 + 11 = 23$
$S(26) + S(27) + S(28) = S(29) \Leftrightarrow 13 + 9 + 7 = 29$
$S(678) + S(679) + S(680) = S(681) \Leftrightarrow 113 + 97 + 17 = 227$
$S(960) + S(961) + S(962) = S(963) \Leftrightarrow 8 + 62 + 37 = 107$

S(3425) + S(3426) + S(3427) = S(3428) ⟺ 137 + 571 + 149 = 857
S(37637) + S(37638) + S(37639) = S(37640) ⟺ 617 + 41 + 283 = 941
S(62628) + S(62629) + S(62630) = S(62631) ⟺ 307 + 389 + 6233 = 6959

slightly more than the number of solutions that exist for the k = 1 case in this range.

The only solution found for the k = 3 case in the range n ≤ 100,000 is

S(63842) + S(63843) + S(63844) + S(63845) = S(63846) ⟺
    233 + 1637 + 1451 + 226 = 3547

Two solutions were found for the k = 4 case in the range n ≤ 100,000

S(1413) + S(1414) + S(1415) + S(1416) + S(1417) = S(1418) ⟺
    157 + 101 + 283 + 59 + 109 = 709

1413 = 3*3*157   1414 = 2*7*101   1415 = 5*283   1416 = 2*2*2*3*59
1417 = 13*109   1418 = 2*709

S(83513) + S(83514) + S(83515) + S(83516) + S(83517) = S(83518) ⟺

    3631 + 449 + 16703 + 20879 + 97 = 41759

83513 = 23*3631   83514 = 2*3*31*449   83515 = 5*16703
83516 = 2*2*20879   83517 = 3*7*41*97   83518 = 2*41759

Only one solution was found for the k = 5 case in the range ≤ 100,000

S(763) + S(764) + S(765) + S(766) + S(767) + S(768) = S(769) ⟺
    109 + 191 + 17 + 383 + 59 + 10 = 769.

Two solutions were found for the k = 6 case in the range ≤ 100,000

S(8786) + S(8787) + S(8788) + S(8789) + S(8790) + S(8791) + S(8792) =
                                                            S(8793)

191 + 101 + 39 + 47 + 293 + 149 + 157 = 977

8786 = 2*23*191   8787 = 3*29*101   8788 = 2*2*13*13*13
8789 = 11*17*47   8790 = 2*3*5*293   8791 = 59*149
8792 = 2*2*2*7*157   8793 = 3*3*977

$$S(42546) + S(42547) + S(42548) + S(42549) + S(42550) + S(42551) + S(42552) =$$
$$S(42553)$$

$$1013 + 271 + 967 + 1091 + 37 + 2503 + 197 = 6079$$

$42546 = 2*3*7*1013 \quad 42547 = 157*271 \quad 42548 = 2*2*11*967$
$42549 = 3*13*1091 \quad 42550 = 2*5*5*23*37 \quad 42551 = 17*2503$
$42552 = 2*2*2*3*3*3*197 \quad 42553 = 7*6079$

And we have more cases where the prime of concern appears once in each of the numbers.

The solution then occurs when there are collections of numbers C_i that make the products a sequence of consecutive integers.

This also gives us a possible explanation for why only one solution was found in both the k=3 and k=5 cases. If all of the numbers are constructed with only one instance of the prime of concern, then we would have the sums

$$\text{odd} + \text{odd} + \text{odd} + \text{odd} \qquad \text{odd} + \text{odd} + \text{odd} + \text{odd} + \text{odd} + \text{odd}$$

which of course must be even. This does not preclude a solution, just forces at least one of the S values to be even.

Conjecture 6: In any range $1 \leq \ldots \leq n$, the number of solutions to a Smarandache p-1 additive sequence will tend to be larger if p is even.

Additional arguments concerning the number of solutions justify the following conjectures.

Conjecture 7: There are an infinite number of positive integers k, such that a solution to the Smarandache k-1 relationship

$$S(n) + S(n+1) + \ldots S(n+k) = S(n+k+1)$$

exists.

Closely related to the previous family of problems, the Smarandache k-1 subtractive relationship would be

$$S(n) - S(n+1) - \ldots - S(n+k) = S(n+k+1)$$

Since the above is algebraically equivalent to

$$S(n) = S(n+1) + S(n+2) + \ldots + S(n+k+1)$$

without deep thought we would expect the number of solutions to be the same as that for the "equivalent" Smarandache k-1 additive relationship.

For the k = 2 case, a search up through n $\leq$ 100,000 yielded the solutions

S(37) - S(38) - S(39) = S(40) $\Leftrightarrow$ 37 - 19 - 13 = 5
37 = 37 38 = 2*19 39 = 3*13 S(40) = 2*2*2*5

S(1383) - S(1384) - S(1385) = S(1386) $\Leftrightarrow$ 461 - 173 - 277 = 11
1383 = 3*461 1384 = 2*2*2173 1385 = 5*277 1386 = 2*3*3*7*11

S(1902) - S(1903) - S(1904) = S(1905) $\Leftrightarrow$ 317 - 173 - 17 = 127
1902 = 2*3*317 1903 = 11*173 1904 = 2*2*2*2*7*17 1905 = 3*5*127

S(4328) - S(4329) - S(4330) = S(4331) $\Leftrightarrow$ 541 - 37 - 433 = 71
4328 = 2*2*2*541 4329 = 3*3*13*37 4330 = 2*5*433 4331 = 61*71

S(4981) - S(4982) - S(4983) = S(4984) $\Leftrightarrow$ 293 - 53 - 151 = 89
4981 = 17*293 4982 = 2*47*53 4983 = 3 * 11*151 4984 = 2*2*3*7*89

S(58970) - S(58971) - S(58972) = S(58973) $\Leftrightarrow$ 5897 - 1787 - 641 = 3469
58970 = 2*5*5897 58971 = 3*11*1787 58972 = 2*2*23*641
58973 = 17*3469

S(91480) - S(91481) - S(91482) = S(91483) $\Leftrightarrow$ 2287 - 227 - 193 = 1867
91480 = 2*2*2*5*2287 91481 = 13*31*227 91482 = 2*3*79*193
91483 = 7*7*1867

For the k = 3 case, a search up through n $\leq$ 100,000 yielded the solutions

S(47) - S(48) - S(49) - S(50) = S(51) $\Leftrightarrow$ 47 - 6 - 14 - 10 = 17
47 = 47 48 = 2*2*2*2*3 49 = 7*7 50 = 2*5*5 51 = 3*17

S(2526) - S(2527) - S(2528) - S(2529) = S(2530) $\Leftrightarrow$ 421 - 38 - 79 - 281 = 23
2526 = 2*3*421 2527 = 7*19*19 2528 = 2*2*2*2*2*79
2529 = 3*3*281 2530 = 2*5*11*23

S(58803) - S(58804) - S(58805) - S(58806) = S(58807) $\Leftrightarrow$
                                                    1153 - 241 - 619 - 22 = 271
58803 = 51*1153 58804 = 2*2*61*241 58805 = 5*19*619
58806 = 2*3*3*3*3*3*11*11 58807 = 7*31*271

49

The number and forms of the solutions justify the following conjecture.

Conjecture 8: There are an infinite number of positive integers k, such that a solution to the Smarandache k-1 relationship

$$S(n) - S(n+1) - S(n+2) - \ldots - S(n+k) = S(n+k+1)$$

exits.

Moving on to another family of similar problems, we now address the case

Problem 6: For what values of k is there a solution to the Smarandache k-k additive relationship

$$S(n) + S(n+1) + \ldots + S(n+k) = S(n+k+1) + S(n+k+2) + \ldots + S(n+k+k+1) ?$$

From previous work, we already know that solutions exist for k = 1 and k = 2 and it has been conjectured that there are an infinite number of solutions to both cases.

A search was performed up through n = 10,000 for the k = 3 case and the following 2 solutions were found

$$S(23) + S(24) + S(25) + S(26) = S(27) + S(28) + S(29) + S(30) \Leftrightarrow$$
$$23 + 4 + 10 + 13 = 9 + 7 + 29 + 5 = 50$$

$$S(643) + S(644) + S(645) + S(646) = S(647) + S(648) + S(649) + S(650) \Leftrightarrow$$
$$643 + 23 + 43 + 19 = 647 + 9 + 59 + 13$$

Once again, one can start with 8 primes such that the sum of 4 equals the sum of the other 4 and use them as the primes of concern for 8 numbers. To obtain a solution, it would then be necessary to find a combination that yields 8 consecutive numbers. As was the case previously, the large number of possibilities can be used to justify the following conjecture.

Conjecture 9 : There are an infinite number of values of k for which the Smarandache k-k additive sequence has a solution.

Of course it is possible to extend the subtractive sequences beyond the k = 1 case already covered. That will not be done here. Furthermore, it is also possible to combine different operators into additional sequences such as

$$S(n) + S(n+1) - S(n+2) + S(n+3) = S(n+4) + S(n+5) - S(n+6) + S(n+7)$$

However, we will not deal with these problems either.

The last problem of tnis form we will consider is the Smarandache 2-1 multiplicative relationship

$$S(n)*S(n+1) = S(n+2)$$

Theorem 26: If the prime of concern in any of the numbers n, n+1 or n+2 is a factor to the first power, then the equation

$$S(n) * S(n+1) = S(n+2)$$

has no solution.

Proof: We know that $S(n) = k_1 p_1$, $S(n+1) = k_2 p_2$ and $S(n+2) = k_3 p_3$ for p_1 a prime factor of n, p_2 a prime factor of n+1 and p_3 a prime factor of n+2. Putting the equation together, we have

$$k_1 p_1 * k_2 p_2 = k_3 p_3$$

Clearly, for this equality to hold, p_1 and p_2 must both divide k_3. From this, it also follows that p_2 must divide k_1 and p_1 must divide k_2. If any of these values is 1, we have a contradiction.

This method of proof can also be used to solve the 2-2-multiplicative problem.

Theorem 27: If the prime of concern of any of the numbers n, n+1, n+2 or n+3 is to the first power, then the equation

$$S(n) * S(n+1) = S(n+2) * S(n+3)$$

has no solution..

Proof: The case where $S(n) = p_1$ will be the only one covered, the others can be dealt with in a similar way.

Special case 1: $n = p_1 = 2$. Then $S(2) = 2$, $S(3) = 3$, $S(4) = 4$ and $S(5) = 5$

Special case 2: $n = p_1 = 3$. Then $S(3) = 3$, $S(4) = 4$, $S(5) = 5$ and $S(6) = 3$.

So we can assume that $p_1 > 3$. Suppose that a solution exists.

Then $S(n+1) = k_2 p_2$, $S(n+2) = k_3 p_3$ and $S(n+3) = k_4 p_4$. Forming the equation,

$$p_1 * k_2 p_2 = k_3 p_3 * k_4 p_4$$

it follows that p_1 must divide either k_3 or k_4. Without loss of generality, assume that p_1 divides k_3. Then

$$S(p_3^k) = k_5 p_1 p_3 \quad \text{where} \quad k \geq k_s p_1$$

Putting all of this together, we have the inequality

$$n+2 \geq p_3^{p_1} \Leftrightarrow n = p_1 \geq p_3^{p_1} - 2$$

which is a contradiction, since $p_3 \geq 2$ and $p_1 \geq 5$.

If the operator is division, we need to carefully differentiate which form of division is being used, integer division or real number division. It is also possible to use the modulus % operator, which is the remainder upon integer division. If integer division is used, we can immediately identify some solutions.

Theorem 28: If p and p+2 are twin primes and the operation is integer division, then

$$\frac{S(p-1)}{S(p)} = \frac{S(p+1)}{S(p+2)}.$$

Proof:
If p is prime, then $S(p-1) / S(p) = 0$, since $S(p-1) \leq p-1 < p$. Therefore, both sides of the equation will evaluate to zero when p is the smaller of a pair of twin primes.

A very large book could be written about Smarandache p-q-{operator name} relationships. Since our purpose here is to introduce the function and some of the consequences, it is time to move on and explore other territory.

In a letter to the editor[10], I. M. Radu posed the question:

For $n > 0$, is it always possible to find a prime p such that $S(n) \leq p \leq S(n+1)$ or $S(n) \geq p \geq S(n+1)$?

This problem was also investigated using the aforementioned C program and it turns out the answer is negative. The program was run for all $n \leq 1,000,000$ and four counterexamples were discovered

$$n = 224 = 2*2*2*2*2*7 \qquad S(n) = 8$$
$$n+1 = 225 = 3*3*5*5 \qquad S(n+1) = 10$$

n = 2057 = 11*11*17	S(n) = 22
n+1 = 2058 = 2*3*7*7*7	S(n+1) = 21
n = 265225 = 5*5*103*103	S(n) = 206
n = 265226 = 2*13*101*101	S(n+1) = 202
n = 843637 = 37*151*151	S(n) = 302
n+1 = 843638 = 2*19*149*149	S(n+1) = 298

The most obvious characteristic of the last two solutions is that each contains two instance of a prime where the primes are twin. Also note that for the solution pair (265225,265226) the products that are not the twin primes are 25 and 26. For the pair (843637,843638) the products are 37 and 38. All of this is easily understood and the explanations point towards additional solutions.

Lemma 9: If the prime of concern in either n or n+1 appears only once, then there is a prime p in the range $S(n) \leq p \leq S(n+1)$ or $S(n) \geq p \geq S(n+1)$.

Proof: Using the definition of prime of concern and theorem 13, either $S(n) = p$ or $S(n+1) = p$ where p is a prime.

Therefore, a potential counterexample to the problem must have a prime of concern that appears at least twice.

Clearly, if one narrows the gap between S(n) and S(n+1), then one increases the chances that there will be no prime between them. If we take two arbitrary numbers m and n where their respective primes of concern are twin primes which appear twice, then $| S(m) - S(n) | = 4$, a gap where only three additional numbers have to be non-prime to satisfy the parameters of the problem. The remaining requirement is then to find the additional factors that do not alter the prime of concern but whose product keeps the gap small. Therefore, the presence of the twin primes is no surprise.

While it is not yet known whether there are an infinite number of twin primes, the following is well-known[11].

For every natural number M, there exists an even number 2k such that there are more than M pairs of successive primes with difference 2k.

The number of potential gaps and the thinning of the primes leads to the conjecture,

Conjecture 10: There are an infinite number of cases where the gap between S(n) and S(n+1) contains no primes. Furthermore, the density of those cases increases as n gets larger.

Rationale: Choose an arbitrary pair of primes p and p+r, and square both

$$p^2 \quad \text{and} \quad p^2 + 2pr + r^2.$$

To satisfy the conditions, we need to form two collections of additional prime factors C_1 and C_2 such that

$$C_1 p^2 \quad \text{and} \quad C_2(p^2 + 2pr + r^2) \quad \text{differ by 1}$$

and p and p+r are the primes of concern for the products. The difference between the respective values of S will be 2r, meaning that an additional 2r-1 numbers must be composite. The range of those numbers would be $2p \leq \ldots \leq 2(p+r)$. As was mentioned before, as the primes p and p+r get larger then the number of possible combinations C_1 and C_2 that keep p and p+r the primes of concern grows at a rapid rate. Also, the values of the product can differ by one in either direction.

The range of testing done by computer $1 \leq n \leq 1,000,000$ is simply not large enough to allow the growth in the number of possibilities to take affect.

Successive values of S(n) can be used to construct a number, and the form of that number was the subject of unsolved problem (8) in [12].

Is r = 0.0234537465 . . . where the sequence of digits is S(n), $n \geq 1$, an irrational number?

Theorem 29: The number r = 0.0234537... formed by the sequence of digits from S(n) is an irrational number.

The following theorem is well-known[13]

Dirichlet's Theorem: If $d > 1$ and a $\neq 0$ are relatively prime, then the arithmetic progression

$$a, a + d, a + 2d, a + 3d, \ldots$$

contains an infinite number of prime numbers.

Proof of theorem 29: Assume that r is rational. Then after some number m of digits, we must have a sequence of k digits that repeat.

$$r = 0.02345 \ldots d_m t_1 t_2 \ldots t_k t_1 t_2 \ldots t_k \ldots$$

Construct the repunit number consisting of 10k 1's.

a = 1111 . . . 1111

and the number

d = 1000 . . . 000

that has 10k zeros.

Since d has only 2 and 5 as prime factors, a and d must be relatively prime. Therefore, by Dirichlet's theorem, the sequence

a, a+d, a+2d, ...

contains an infinite number of primes.

Since S(p) = p and we have a prime with more digits than the number k that repeat, all the digits of the repeated segment of r must be 1's.

Now, construct another number a that consists of 10n 3's

a = 3333. . . . 3

and use the same value for d. Again, a and d are relatively prime, so the sequence

a, a + d, a + 2d, . . .

contains an infinite number of primes.

We are now forced into the contradiction of having all of the digits of the repeated segment being 1 and having some of them equal to 3. Therefore, the number r does not repeat and must be irrational.

We have already proven that $0 \leq S(n) / n \leq 1$ where S(1) / 1 is the only case where the value is zero and S(p)/p = 1 for p a prime. The topic of unsolved problem (7) in [13] concerned what happens away from the endpoints.

Are the points S(n) / n uniformly distributed in the interval (0,1)?

The answer to this question is no. However here we will prove a much stronger result.

Theorem 30: There is no interval (0,b), b < 1 and b a real number, where the points

S(n) / n

are evenly distributed.

Lemma 10: For any real $\epsilon > 0$, there is some number M such that

$S(n) / n < \epsilon$ for all $n \geq M$ and n not prime

Proof: Let p be a prime such that $2 / p < \epsilon$. Then

$$\frac{S(p^2)}{p^2} = \frac{2p}{p^2} = \frac{2}{p} < \epsilon .$$

By theorem 11, it follows that

$$\frac{S(p^k)}{p^k} < \epsilon \quad \text{for } k \geq 2.$$

Using the ordering properties of the integers, it is possible to find the smallest prime P such that $2 / P < \epsilon$.

Now, let p be an arbitrary prime less than P. By theorem 11, there is some number K such that

$$\frac{S(p^k)}{p^k} < \epsilon \quad \text{for all } k \geq K$$

For each prime $p < P$, find the corresponding K such that

$$\frac{S(p^K)}{p^K} < \epsilon$$

and use those numbers to create a list L. Since this list is finite, there is no need to involve infinite processes.

If we then set

$$M = \max \{ L \cup P \}$$

we have a value of M satisfying the criteria of the lemma.

Proof of theorem 30: By the previous lemma, for any real number $0 < a < 1$, the set

$\{ n \mid S(n) / n > a \text{ and n not prime} \}$

is finite. From this, it follows that there is no interval (0,b) where the points are evenly distributed.

Clearly, S(n) / n is always rational and the set

$$\{\, r \mid r = S(n) / n \text{ for some } n \geq 1 \,\}$$

infinite. Therefore, the number of solutions is countable. With this idea and the results of the previous theorem, we can ask the question

For what rational numbers r does there exist an n such that r = S(n) / n ?

One result is quite easy. Let

$$Q = \{\, p \mid \text{there exists some } n \text{ such that } S(n) / n = 1 / p \,\}$$

Theorem 31: $Q = \{\, 2,3 \,\}$

Proof:

$$S(8) = 4, \text{ so } S(8) / 8 = 1/2$$
$$S(27) = 9, \text{ so } S(27) / 27 = 1/3$$

Let $p \geq 5$ and prime. If we take the successive powers of that prime and form the ratios, we have

$$\frac{S(p^2)}{p^2} = \frac{2p}{p^2} = \frac{2}{p} \qquad \frac{S(p^3)}{p^3} = \frac{3p}{p^3} = \frac{3}{p} \qquad \frac{S(p^4)}{p^4} = \frac{4p}{p^4} = \frac{4}{p}$$

on up to

$$\frac{S(p^{p-1})}{p^{p-1}} = \frac{p-1}{p^{p-2}} \qquad \frac{S(p^p)}{p^p} = \frac{p*p}{p^p} = \frac{1}{p^{p-2}}$$

Which has the form $1 / p$ only when $p = 3$. Since

$$\frac{1}{p} > \frac{1}{p^{p-2}} = \frac{S(p^p)}{p^p} > \frac{S(p^{p+1})}{p^{p+1}}$$

it follows that no power of p can have the form $1 / p$.

It should also be clear that there is no power of another prime q such that

$$\frac{S(q^k)}{q^k} = \frac{1}{p}$$

For suppose that n has more than one prime factor and p_i is the prime of concern with multiplicity j. Then

$$\frac{S(n)}{n} = \frac{kp_i}{n} < \frac{S(p_i^j)}{p_i^j}$$

and the ratio cannot be of the form 1 / p for any prime p.

We will close the study of the Smarandache function by briefly touching on several additional problems.

Unsolved problem (21) in [14] deals with products.

Are there m,n,k non-null positive integers, $m \neq 1 \neq n$, for which

$$S(m*n) = m^k * S(n)?$$ Clearly, S is not homogenous to degree k.

Clearly m = n = 2, k = 1 is a solution. The search for other solutions, if any, will be left for others to explore.

Unsolved problem (22) in [15] deals with logarithms.

Is it possible to find two distinct numbers k,n for which

$$\log_{S(k^n)} S(n^k)$$

is an integer?

The answer is yes, k = 4, n = 2 is a solution. In fact, all numbers a and b where

$$a^b = b^a$$

will be a solution.

Mersenne numbers are of the form $2^m - 1$. It is known that some are prime and some are composite. Clearly, if a Mersenne number p is prime, then

$$S(p) = p$$

and we have a pair of Mersenne primes that satisfies the Smarandache relation. The next question is then

Is there a pair of Mersenne numbers (m,n) such that m is composite and

$$S(m) = n$$

The answer is affirmative, since $2^6 - 1 = 63$ $S(63) = 7$ and $2^3 - 1 = 7$. Whether other such solutions exist is a question that will be left open.

Unsolved problem (2094) in [16] deals with greatest common divisor, gcd.

Solve the diophantine equation

$$gcd(x,y) = gcd(S(x),S(y))$$

If we take p,q to be distinct primes greater than 2, then

$$gcd(2p^2,2q^2) = 2 \qquad S(2p^2) = 2p \quad S(2q^2) = 2q \quad and$$

$$gcd(2p,2q) = 2$$

so there are an infinite number of solutions to the equation. In fact, it is a simple matter to construct several such infinite families.

And now, we come to the author imposed end of our journey through only a tiny part of the ramifications of the Smarandache function. It is my sincere hope that you, the reader, feel enriched, yet hungry for more. If you do, then I have received the highest possible praise that an author can receive. Happy with the product, yet eager for more. Good luck in your mathematical endeavors.

Bibliography

1. F. Smarandache, A Function in the Number Theory, An. Univ. Timisoara, seria st. mat., Vol. XVIII, fasc. 1, pp. 79-88, 1980; **Mathematical Reviews**: 83c: 10008.

2. R. Muller, 1993. **Unsolved Problems Related to Smarandache Function**. Phoenix: Number Theory Publishing Company.

3. R. Muller ed. **Smarandache Function Journal**. Phoenix: Number Theory Publishing Company.

4. Richard K. Guy, 1994. **Unsolved Problems in Number Theory: Second Edition**. New York: Springer-Verlag.

5. R. Muller, 1993. **Unsolved Problems Related to Smarandache Function**. Phoenix: Number Theory Publishing Company.

6. T. Yau, 1994. A Problem Concerning the Fibonacci Series. **Smarandache Function Journal** Vol. 4-5, No. 1: 42.

7. Paulo Ribenboim, 1989. **The Book of Prime Number Records: Second Edition**. New York: Springer-Verlag.

8. Richard K. Guy, 1994. **Unsolved Problems in Number Theory: Second Edition**. New York: Springer-Verlag.

9. Jorge Castillo, 1995. Unpublished manuscript.

10. I. M. Radu, Letter to the Editor. **Mathematical Spectrum** Vol. 27, No. 1.

11. Paulo Ribenboim, 1989. **The Book of Prime Number Records: Second Edition**. New York: Springer-Verlag.

12. R. Muller, 1993. **Unsolved Problems Related to Smarandache Function**. Phoenix: Number Theory Publishing Company.

13. Ibid.

14. Ibid.

15. Ibid.